FLORA OF TROPICAL EAST AFRICA

SCYTOPETALACEAE

B. VERDCOURT

Trees or shrubs, sometimes cauliflorous. Leaves simple, exstipulate, alternate, sessile or petiolate, entire or very rarely denticulate. Inflorescences axillary, terminal or often borne on the old stems, paniculate, racemose or in fascicles; bracts small, soon falling. Flowers hermaphrodite, pedicellate, the pedicel sometimes jointed at the apex. Calyx cupuliform or shallowly tray-shaped, entire, lobed or slightly toothed, persistent at the base of the fruit. Petals 3–16, valvate or sometimes the bud bursting somewhat irregularly; lobes free or joined at the base. Stamens numerous, in 3–6 ranks inserted on the annular disk, essentially epipetalous, often falling with the corolla; filaments free or joined at the base; anthers basifixed. Ovary superior or inferior, 3–8-locular, each locule with 2–several ovules; style 1, filiform; stigma small, entire or slightly lobed. Fruit a woody or crustaceous capsule, or indehiscent but scarcely fleshy, unilocular. Seeds 1–8, with abundant endosperm which is often ruminate.

A small family of 5 genera and about 20 species confined to tropical Africa; its affinities almost certainly lie with *Olacaceae* rather than *Tiliaceae* as is often claimed.

BRAZZEIA

Baill. in Bull. Soc. Linn. Paris 1: 609 (1886); Letouzey in Adansonia 1: 129 (1961)

Erytropyxis Pierre in Bull. Soc. Linn. Paris 2: 1265 (1896)

Pseudobrazzeia Engl., V.E. 3(2): 473 (1921)

Shrubs or small trees. Leaves subsessile, membranous or coriaceous, entire, undulate or denticulate. Flowers solitary or in fascicles on the old wood; pedicels not articulated. Calyx entire or lobed. Corolla of 2–5 petals or irregularly divided, often falling as one piece. Filaments of stamens free; anthers short, dehiscing by an apical pore. Ovary 4–8-locular, each locule with several ovules; style glabrous; stigma subcapitate, entire or slightly lobed. Capsules globular or almost so, usually eventually dehiscent, but sometimes fruit apparently indehiscent, many-seeded. Endosperm not ruminate.

A genus of 3 species one of which occurs in the extreme western part of the Flora area.

B. longipedicellata *Verdc.* in K.B. 5: 344, fig. 2 (1951); I.T.U., ed. 2: 404 (1952); Letouzey in Adansonia 1: 132 (1961); Germain in F.C.B. 10: 323 (1963); Verdc. in K.B. 17: 500 (1964). Type: Uganda, Kigezi District, Ishasha Gorge, *Purseglove* 2002 (K, holo.!, EA, iso.!)

Small glabrous evergreen tree 4·5–8(–12) m. tall. Leaf-lamina subcoriaceous, elliptic to oblong-elliptic, (2·3–)4–13 cm. long, (1·3–)2–5·5 cm. broad, acuminate, cuneate; petiole short and thick, ± 3 mm. long. Flowers

FIG. 1. *BRAZZEIA LONGIPEDICELLATA.* **1,** leafy branch, × $\frac{2}{3}$; **2,** flower-buds, × 1; **3,** flower, × 1; **4,** stamen, × 4; **5,** upper part of gynoecium, × 4; **6,** fruit, × 1; **7,** cross section of fruit, × 1; **8,** seed-mass, × 1; **9,** seed, × 1. 1–5, from *Purseglove* 2002; 6–9, from *Purseglove* 3059.

fleshy, in clusters on the old wood, each cluster being made up of numerous 3-flowered fascicles; pedicels 1·5–4 cm. long. Calyx cupular, 12 mm. in diameter, closely adpressed to the corolla in bud, at first ± entire or irregularly 3–4-lobed but later tearing into a number (7 or so) of squarish irregular lobes. Corolla white or rose, rather fleshy, conical in bud, dividing into 3–4 lobes or tearing irregularly into 3–4 subequal or unequal parts, with ribbed interior surfaces and irregular margins, finally curving back and falling off with the stamens attached, ± 3 cm. across. Stamens very numerous, ± 6 mm. long. Ovary globose, 4 mm. in diameter; style 9–10 mm. long. Fruit woody, brown, globose, ± 2·8–3 cm. in diameter, 6-ribbed but probably not dehiscent. Seeds ± 8, adhering in a globose mass, each elliptic, compressed, 9 mm. long, 7 mm. wide, 4·5 mm. thick, with an elliptic groove above and below which separates a smooth outer annular area from the central raised portions. Fig. 1.

UGANDA. Kigezi District: Ishasha Gorge, 10 Feb. 1945, *Greenway & Eggeling* 7105! & Mar. 1946, *Purseglove* 2002! & Aug. 1949, *Purseglove* 3059! & *Purseglove* 3111*

DISTR. U2; Congo Republic

HAB. Dense rain-forest, riverine forest of *Allanblackia*, *Chrysophyllum*, *Newtonia* and *Parinari* in rocky gorges; 1200–1500 m. (but recorded down to 850 m. in Congo)

* Fide I.T.U., ed. 2; at Kew this sheet is a *Solanum*.

INDEX TO SCYTOPETALACEAE